CREATIVE TEACHING ASSOCIATES

D O M I N O

BOOK A

by ARTHUR WIEBE

The author and publisher hereby grant permission to teachers to duplicate these pages for classroom use.

Copyright © 1973
CREATIVE TEACHING ASSOCIATES
P.O. Box 7766, Fresno, California 93747
Printed in the United States of America

DOMINO MATH   Book A

The Domino Math series provides abundant opportunity for the student to engage in problem solving activities employing dominoes. Inherent to these exercises is extensive drill in basic addition and subtraction facts in a mode that is challenging and interesting.

Each activity begins at either the manipulative or representational level. The student makes the interpretation into the abstract. All exercises provide for several ways of self-checking so that the child receives immediate reinforcement for correct responses. This saves the teacher from the need to monitor or correct all of the work in the usual way. The child develops a sense of self-reliance and develops in problem solving ability.

The materials have been thoroughly classroom tested and have received the endorsement of participating teachers. They should provide a useful addition to your inventory of problem types whose objective is mastery of basic facts.

It is suggested that each classroom have several sets of regular size double-nine sets of dominoes. The double-nine set is specified since it contains all the basic combinations. Regular, rather than club, dominoes are recommended because all of the forms are drawn to fit this size and because they are less expensive. Each set of double-nine dominoes contains fifty-five dominoes. So, several groups of students can be accomodated by a set.

The teacher purchasing this volume is hereby granted the restricted right to reproduce copies of these lessons for use in the classroom. The permission to reproduce is limited for use in a single classroom.

Book A contains addition and subtraction problems. Book B focuses on multiplication and division problems.

One of the great advantages of domino mathematics is that the student is constantly translating the real world into symbols. This will lead the child to the understanding that symbols tell him something about the real world. Through repeated experience, the child becomes keenly aware of the fact that mathematics can be used to describe his world. Once convinced of this fact, the student becomes a more highly motivated and effective learner.

There are just two ways in which we can translate the real world into abstract symbols: by counting and by measuring. In this book, counting is emphasized.

There is constant emphasis on counting. Counting is a basic skill and students should engage in it frequently. But counting is a tedious and time consuming way to arrive at a result. So, we want to present the child with experiences that will make him impatient with counting and ready for shortcuts so that he can be weaned from counting. In these activities, the first step away from counting is the recognition of patterns. Five dots are soon associated with four on the four corners of an imaginary square and one in the center. Six dots are two rows of three. A second shortcut that soon gains interest is that of knowing the basic addition facts. This often eliminates the need for counting to arrive at the sum. The mastery of basic addition facts becomes a way to save time and effort and that is acceptable to nearly all of us. Such a reason has much more appeal than being told that mastery of facts is essential for later mathematics. All of us like to achieve something with minimum effort. Here, this desire is translated into the reason for knowing the addition facts.

So, inward motivation builds up for mastering these basic facts. Later, multiplication will be shown to be another labor-saving device.

But counting still retains a valid function. In a sense, it is a back-up system for checking answers. It serves as the primary skill upon which the student can depend as he develops new skills.

Another advantage that will soon become apparent in the use of dominoes is that subsets can be selected to control the difficulty level. In early work, for example, only dominoes with one, two, or three dots on a half can be used. As children master these, new ones may be added on a selective basis.

The following are specific comments about the major sections in the book.

Pages 1 and 2 contain a master set of domino pictures suitable for reproduction. The reproductions can be used in several ways. They can be cut out and pasted on the blank masters to create additional lessons. They can be mounted on tag board, cut out, and used in place of wooden dominoes. Each student can have such a paper set at little cost.

Pages 3 and 4 are useful for various classification exercises. For example, page 3 can be cut into three strips. The strips pasted end to end, and then cut so that 19 spaces remain. These are then labeled from 0 to 18. The student takes the double-nine set and places all dominoes with no dots on the first space, those with a total of one dot on the second space, etc. When all have been properly classified, which means the student has worked fifty-five addition problems, they will form a symmetric arrangement. This symmetry is a quick way to check all of the results.

Another way to stack dominoes is by differences. The doubles have a difference of zero and are all stacked on the second space, etc. The last space will have the 9-0 domino, which is the only one with a difference of 9. This will form a staircase coming down. Again, the arrangement is the self-correcting feature.

Dominoes can also be stacked by remainders. Divide the smaller number into the larger and stack by remainder.

A fourth activity is to place all dominoes with a half that has no dots on the first stack; after that, place any remaining ones with a single dot on a half on the second stack, etc. This yields another pattern. Of course, the exercise can be reversed by first stacking those with nine dots on a half, etc.

Page 5 is a blank form. Such blank forms introduce each type of problem in the book and they can be used to reproduce copies and then write additional lessons. Students may be encouraged to write lessons and exchange them. There is something highly motivating about mathematics that is created in the classroom. For a more professional result, the dominoes from the first pages may be reproduced and mounted on a master for reproduction.

Page six is restricted to simple combinations of numbers from one to five. The child may approach the first example, which is solved in several ways. First, the number of dots in each half is counted and recorded. This is a translation from the real world to symbols when dominoes are used, or from the representational form to symbols when this sheet is used. If the child knows that $1 + 2 = 3$, then the answer is simply recorded. If this is not known, the child does not need to wait in frustration. Counting can always be relied on and so the total number of dots is counted and recorded. Counting can also be used to check on the answer given from knowing that $1 + 2 = 3$. Thus, these two ways of arriving at the answer means that most children can find it and then use one to check the other.

Pages 7 and 8 continue with this type of problem but with more difficult combinations.

On pages 9 and 10, problem solving experiences in which each problem will have several answers are introduced. The student will examine many dominoes in finding those that fit. Each rejected domino as well as each accepted domino is drill in basic addition facts. A domino can be rejected correctly only if an estimate or actual count shows that it contains the wrong number of dots, or if the sum found by counting those in each half and adding does not match the requirement. It is likely that students will work six or more problems for each domino selected and thus this page may represent work on fifty or more problems.

Pages 12 to 14 introduce more difficult examples with addends given. Pages 15 and 16 are problems with only the sums given.

Beginning with the blank form on page 17, subtraction is introduced. This form is for writing additional original problems, either by the students or the teacher. Students can formulate problems of their own and exchange them.

On pages 18 to 20 you will find subtraction problems with the answers missing. On pages 21 to 23 only the answer is given and the other elements are to be found. Here again, the child may make many trials before reaching a solution and there will be several solutions. Beginning with the blank form on page 24 through page 27, a new form is introduced that intensifies the computation required. The number of dots in each domino is determined and recorded. This can be done by the basic process of counting or the more advanced method of recognition by patterns and adding. Then, the horizontal addition is performed. The answer can be checked by counting all of the dots. Three problems, two vertical and one horizontal, are checked by one counting. There is absolutely no need for the child to get stuck and frustrated because no one is available to help. Counting gets him unstuck.

On page 28, the student has great freedom in the choice of dominoes that fit the stated facts. There will be a variety of answers. In fact, an independent investigation suggests itself: determine all possible answers in each case. The change in format from section to section continues the basic drill and practice without introducing too great familiarity so that interest is lost.

Beginning with page 29, the student will have to subtract and add. In each case there are two subtraction problems and one addition problem. All can be checked with one counting of the dots.

On page 31, the missing addend must be found in the abstract form. Then the addends are used to complete the dominoes. In some cases there is just one answer as in the second domino of the first problem. But the first domino can be completed in several ways. In the end, all the work can be checked by counting and comparing with the sum in the square.

Page 32 opens up the situation still more. There are several paths to the solution and there are numerous correct answers.

Page 33 varies the format again and here, too, there will be several ways to complete each problem.

On page 35 the student is asked not only to find the vertical sums, but also the horizontal sum of dots. First, the number of dots in each separate domino is found and recorded along the bottom. Next the sum of dots in the top and bottom halves are found and recorded to the right. The square contains the sum found by adding the horizontal sums and the vertical sums. Since this is arrived at in two ways, there is a check. Beyond that, there is again the basic check of counting all the dots. By this time, the student will probably be comfortable in checking through adding in two ways rather than relying on counting.

Beginning with page 36, strategy requirements increase. In the third example, for instance, the third domino is completed first, then the top of the first domino. 31-13 yields the fact that the number of dots to be placed in bottom halves is 18. But 9 are already in place so the remaining 9 must be divided between the first two dominoes. This can be done in several ways.

Page 37 can be approached in several ways. It is suggested that it be tried by the teacher before it is presented to the class. Students may be challenged to find the most unusual way in which it can be used.

The form on page 39 can be used if more advanced problems are of particular interest to certain students with greater ability. Problems of any of the previous types can be formulated.

On page 40, the mainstream of development continues with easier problems in which horizontal and vertical sums must be found. Each of the problems through page 44 require the solution to six problems. The final two additions serve as a check on each other. Counting all the dots is still the backup check!

Pages 44 to 47 also contain six problems in each example but they are a mix of addition and subtraction problems.

Pages 47 and 48 will have many answers for each problem. Some students may want to investigate all the possible answers for a given problem.

Beginning with the blank form on page 49 and continuing through page 51 another dimension is added. Not only are the horizontal and vertical sums to be found, but also the diagonal sums. The upper half of the left domino is added to the lower half of the right domino, and the sum is recorded in the left circle. The other diagonal is used to find the sum in the right circle. The sum of the numbers named in the circles should be the same as that in the square. The sum in the square is found and checked in four ways: by horizontal, vertical, and diagonal addition, and by counting. Each example consists of nine addition problems. All nine problems can be checked by counting once. Thus, page 50 really contains 54 problems.

Pages 52 and 53 require subtraction in addition to finding the sum.

Page 55 introduces another type of problem. It can be used in conjunction with any of the possible sums entered in the center. Specific problems are shown on pages 56 and 57. On page 56, the top row, the bottom row, the left column, and the right column should all just have six dots. The child may choose from the full set. In the second case the requirement is the same, except that the sum is nine dots on each side. Do not count the center space with the circle. It is just there to state the target sum.

These pages require massive drill and many solutions which will not fit. But even these serve our purpose of having the student drill in fundamentals.

On page 57 part of the solution has been given. Subtraction and addition and then finding the dominoes will be required to complete the problems.

An idea of the amount of drill that is possible with these problems can be gained from the following example. Suppose we give the student nine dominoes to work with. Four of these will ultimately be used in the solution, but which four is up to the student. Any number can be the target sum and it can change as the student desires. At least 30,000 different combinations are possible and several of these will be correct. In fact, hundreds of them may be. But to be able to say there is no solution when starting with nine dominoes would require working 30,000 or more separate addition problems.

The solutions are easier as the number of dominoes available for use is increased and more difficult as the number is decreased.

Page 58 extends this same type of problem to four halves in each direction. The rules are the same.

Pages 59 and 60 can be used to provide addition drill in another format. A stated number of dominoes can be used or the full set may be used to find four sets of four on page 59, or three sets of three on page 60 that have the same sum. Reducing the number of available dominoes makes the solutions more difficult to find.

© 1973 Creative Teaching Associates

© 1973 Creative Teaching Associates

© 1973 Creative Teaching Associates

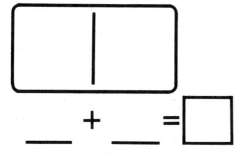

___ + ___ = ☐

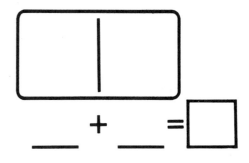

___ + ___ = ☐

___ + ___ = ☐

___ + ___ = ☐

___ + ___ = ☐

___ + ___ = ☐

___ + ___ = ☐

___ + ___ = ☐

___ + ___ = ☐

___ + ___ = ☐

© 1973 Creative Teaching Associates

FIND THE SUM.

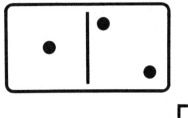

_1_ + _2_ = 3

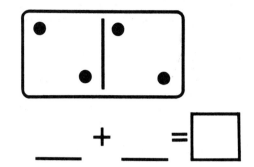

__ + __ = ☐

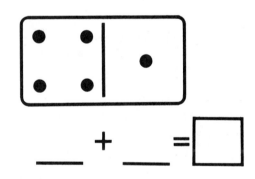

__ + __ = ☐

__ + __ = ☐

__ + __ = ☐

__ + __ = ☐

__ + __ = ☐

__ + __ = ☐

© 1973 Creative Teaching Associates

FIND THE SUM.

NAME _____ ⑦

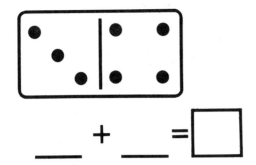

___ + ___ = ☐

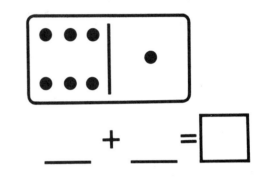

___ + ___ = ☐

___ + ___ = ☐

___ + ___ = ☐

___ + ___ = ☐

___ + ___ = ☐

___ + ___ = ☐

___ + ___ = ☐

© 1973 Creative Teaching Associates

FIND THE SUM.

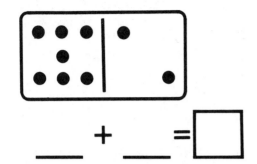

___ + ___ = ☐

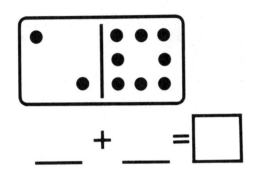

___ + ___ = ☐

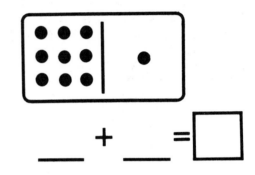

___ + ___ = ☐

___ + ___ = ☐

___ + ___ = ☐

___ + ___ = ☐

___ + ___ = ☐

© 1973 Creative Teaching Associates

FIND A DIFFERENT DOMINO FOR EACH CASE.

__ + __ = 6

__ + __ = 7

__ + __ = 6

__ + __ = 8

__ + __ = 7

__ + __ = 8

__ + __ = 7

__ + __ = 8

© 1973 Creative Teaching Associates

FIND A DIFFERENT DOMINO FOR EACH CASE.

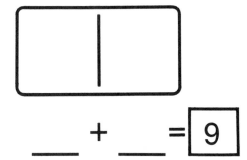

__ + __ = ⬜ 9

__ + __ = ⬜ 10

__ + __ = ⬜ 9

__ + __ = ⬜ 10

__ + __ = ⬜ 9

__ + __ = ⬜ 10

__ + __ = ⬜ 9

__ + __ = ⬜ 10

© 1973 Creative Teaching Associates

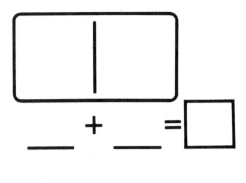

___ + ___ = ☐

___ + ___ = ☐

___ + ___ = ☐

___ + ___ = ☐

___ + ___ = ☐

___ + ___ = ☐

___ + ___ = ☐

___ + ___ = ☐

___ + ___ = ☐

___ + ___ = ☐

© 1973 Creative Teaching Associates

FIND THE SUM.

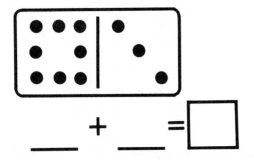

__ + __ = ☐

__ + __ = ☐

__ + __ = ☐

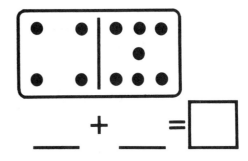

__ + __ = ☐

__ + __ = ☐

__ + __ = ☐

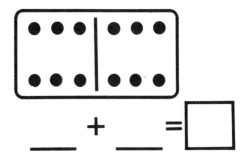

__ + __ = ☐

__ + __ = ☐

__ + __ = ☐

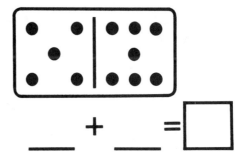

__ + __ = ☐

© 1973 Creative Teaching Associates

FIND THE SUM.

___ + ___ = ☐

___ + ___ = ☐

___ + ___ = ☐

___ + ___ = ☐

___ + ___ = ☐

___ + ___ = ☐

___ + ___ = ☐

___ + ___ = ☐

___ + ___ = ☐

___ + ___ = ☐

© 1973 Creative Teaching Associates

FIND THE SUM.

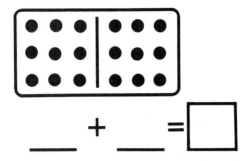

___ + ___ = ☐

___ + ___ = ☐

___ + ___ = ☐

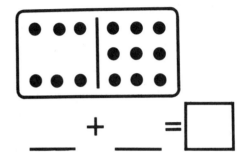

___ + ___ = ☐

___ + ___ = ☐

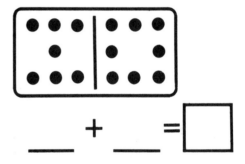

___ + ___ = ☐

___ + ___ = ☐

___ + ___ = ☐

___ + ___ = ☐

___ + ___ = ☐

© 1973 Creative Teaching Associates

NAME _____

FIND A DIFFERENT DOMINO FOR EACH.

___ + ___ = 11

___ + ___ = 12

___ + ___ = 11

___ + ___ = 12

___ + ___ = 11

___ + ___ = 12

___ + ___ = 11

___ + ___ = 13

___ + ___ = 12

___ + ___ = 13

© 1973 Creative Teaching Associates

FIND A DIFFERENT DOMINO FOR EACH.

__ + __ = 18

__ + __ = 15

__ + __ = 17

__ + __ = 14

__ + __ = 16

__ + __ = 14

__ + __ = 16

__ + __ = 14

__ + __ = 15

__ + __ = 13

© 1973 Creative Teaching Associates

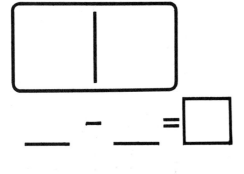

__ __ − __ __ = ☐

__ __ − __ __ = ☐

__ __ − __ __ = ☐

__ __ − __ __ = ☐

__ __ − __ __ = ☐

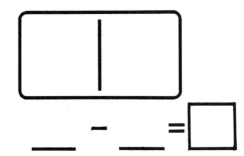

__ __ − __ __ = ☐

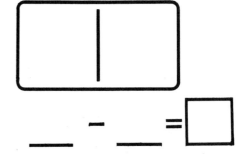

__ __ − __ __ = ☐

__ __ − __ __ = ☐

__ __ − __ __ = ☐

__ __ − __ __ = ☐

© 1973 Creative Teaching Associates

FIND THE DIFFERENCE.

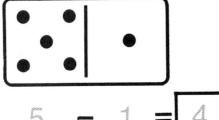

_5_ – _1_ = 4

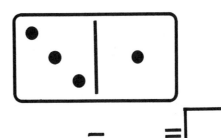

___ – ___ = ☐

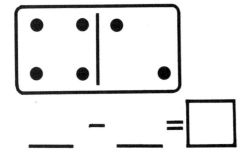

___ – ___ = ☐

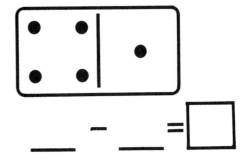

___ – ___ = ☐

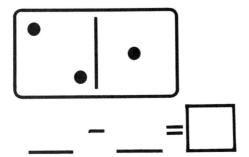

___ – ___ = ☐

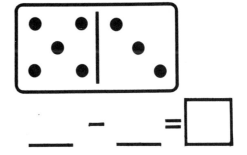

___ – ___ = ☐

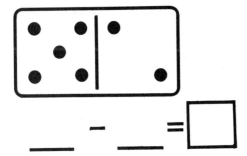

___ – ___ = ☐

___ – ___ = ☐

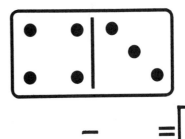

___ – ___ = ☐

© 1973 Creative Teaching Associates

FIND THE DIFFERENCE.

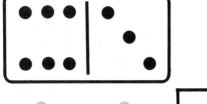

6   –   3   = ☐

___ – ___ = ☐

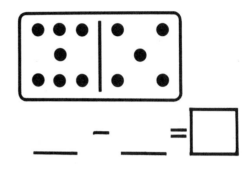

___ – ___ = ☐

___ – ___ = ☐

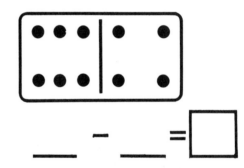

___ – ___ = ☐

___ – ___ = ☐

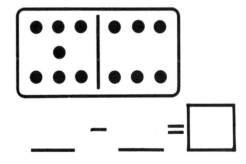

___ – ___ = ☐

___ – ___ = ☐

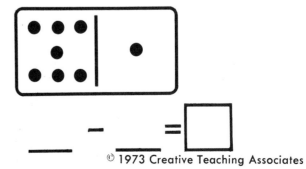

___ – ___ = ☐

© 1973 Creative Teaching Associates

FIND THE DIFFERENCE.

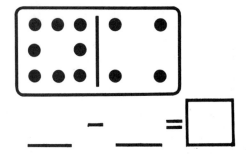

_____ − _____ = ☐

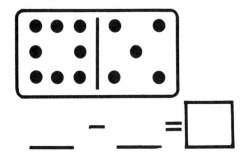

_____ − _____ = ☐

_____ − _____ = ☐

_____ − _____ = ☐

_____ − _____ = ☐

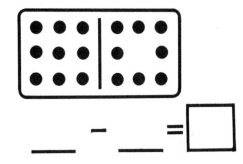

_____ − _____ = ☐

_____ − _____ = ☐

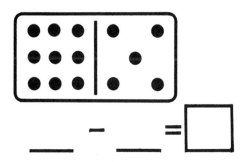

_____ − _____ = ☐

_____ − _____ = ☐

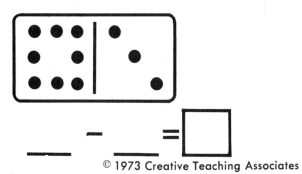

_____ − _____ = ☐

© 1973 Creative Teaching Associates

FIND A DIFFERENT DOMINO FOR EACH.

___ – ___ = 2

___ – ___ = 3

___ – ___ = 2

___ – ___ = 3

___ – ___ = 2

___ – ___ = 3

___ – ___ = 2

___ – ___ = 3

___ – ___ = 2

___ – ___ = 3

© 1973 Creative Teaching Associates

FIND A DIFFERENT DOMINO FOR EACH.

___ − ___ = 4

___ − ___ = 5

___ − ___ = 4

___ − ___ = 5

___ − ___ = 4

___ − ___ = 5

___ − ___ = 4

___ − ___ = 5

___ − ___ = 4

___ − ___ = 5

© 1973 Creative Teaching Associates

NAME _____ 23

FIND A DIFFERENT DOMINO FOR EACH.

___ − ___ = 6

___ − ___ = 7

___ − ___ = 6

___ − ___ = 7

___ − ___ = 6

___ − ___ = 8

___ − ___ = 6

___ − ___ = 8

___ − ___ = 7

___ − ___ = 9

© 1973 Creative Teaching Associates

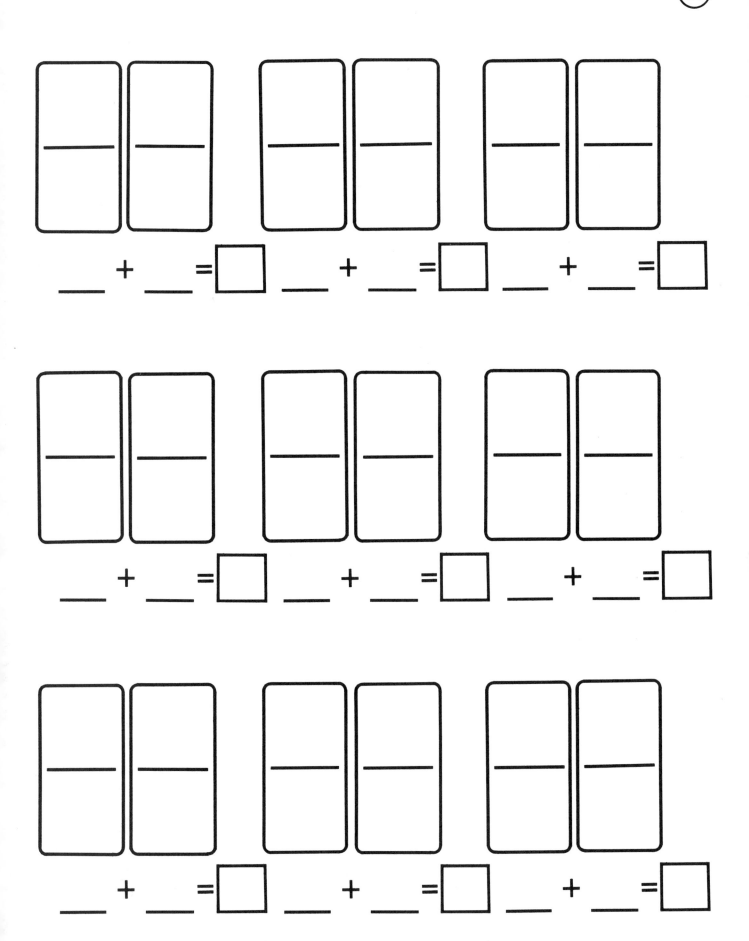

© 1973 Creative Teaching Associates

FIND THE SUM.

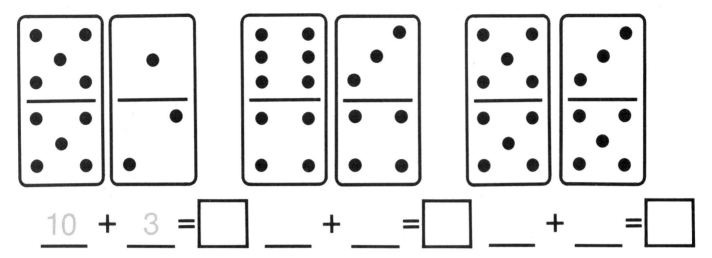

10 + 3 = ☐   ___ + ___ = ☐   ___ + ___ = ☐

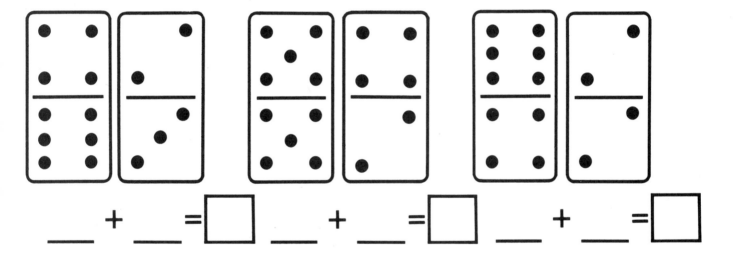

___ + ___ = ☐   ___ + ___ = ☐   ___ + ___ = ☐

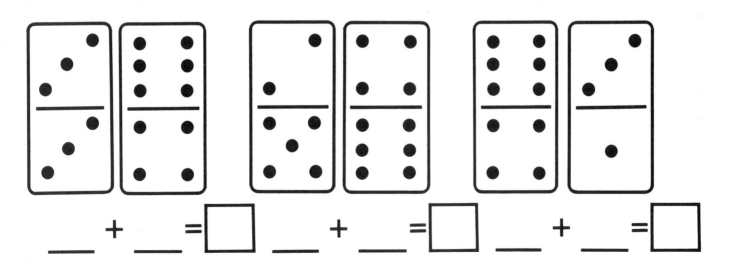

___ + ___ = ☐   ___ + ___ = ☐   ___ + ___ = ☐

© 1973 Creative Teaching Associates

NAME_____ 26

FIND THE SUM.

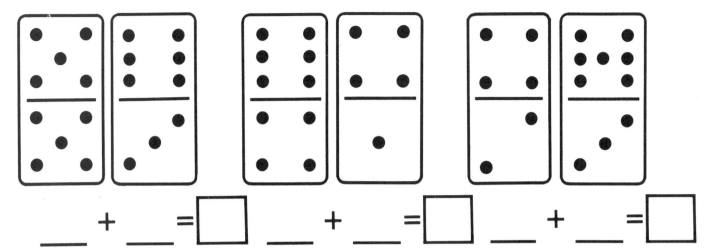

___ + ___ = ☐   ___ + ___ = ☐   ___ + ___ = ☐

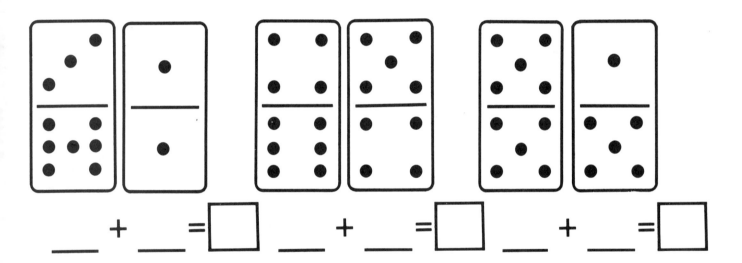

___ + ___ = ☐   ___ + ___ = ☐   ___ + ___ = ☐

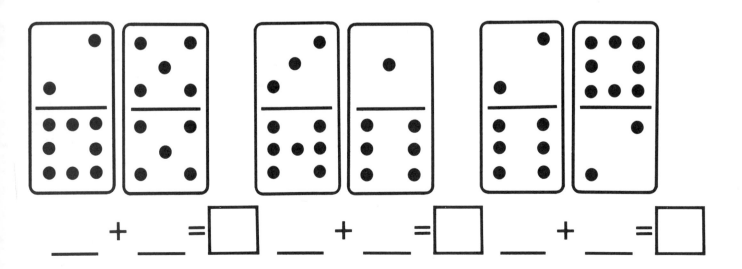

___ + ___ = ☐   ___ + ___ = ☐   ___ + ___ = ☐

© 1973 Creative Teaching Associates

FIND THE SUM.

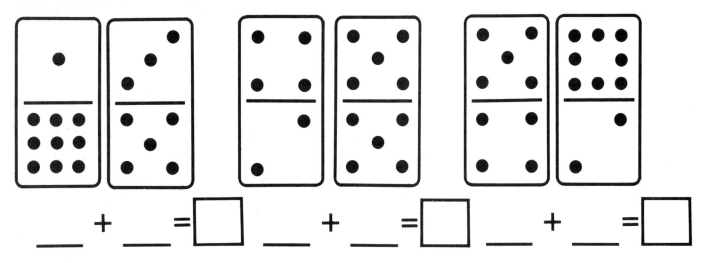

___ + ___ = ☐   ___ + ___ = ☐   ___ + ___ = ☐

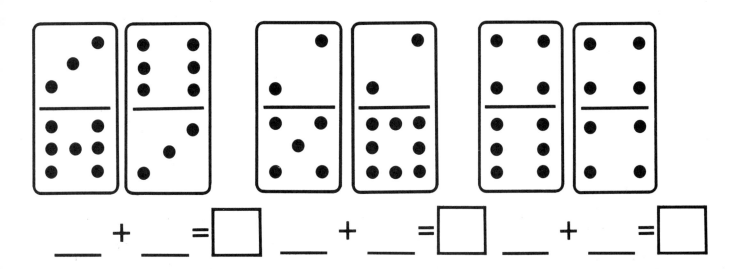

___ + ___ = ☐   ___ + ___ = ☐   ___ + ___ = ☐

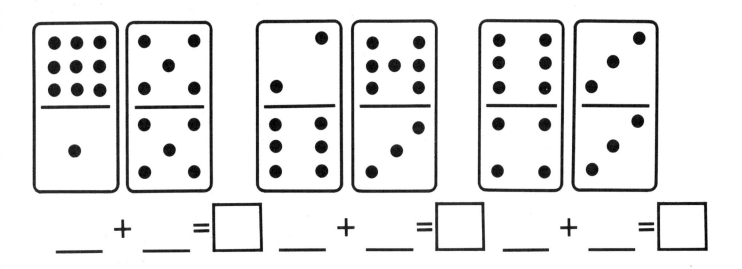

___ + ___ = ☐   ___ + ___ = ☐   ___ + ___ = ☐

© 1973 Creative Teaching Associates

FIND DOMINOES FOR EACH PROBLEM.   THEN FIND THE SUM.

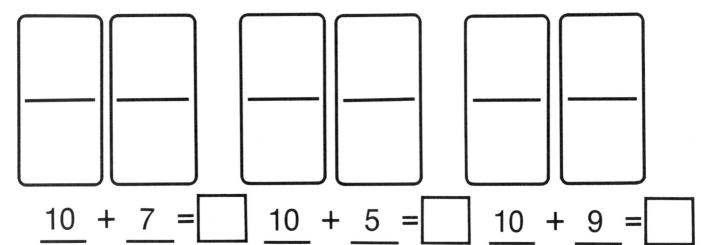

10 + 7 = [ ]   10 + 5 = [ ]   10 + 9 = [ ]

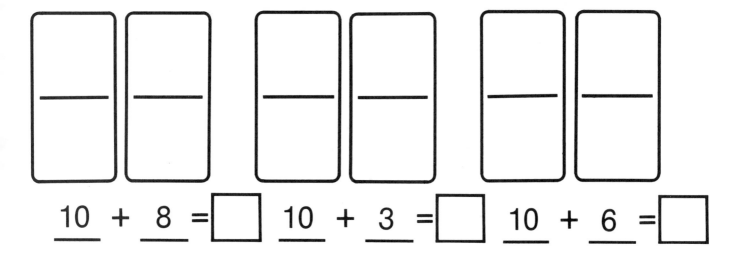

10 + 8 = [ ]   10 + 3 = [ ]   10 + 6 = [ ]

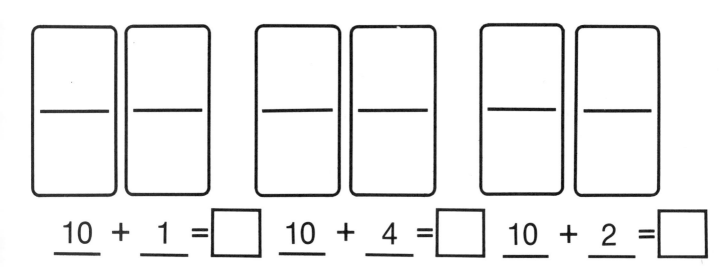

10 + 1 = [ ]   10 + 4 = [ ]   10 + 2 = [ ]

© 1973 Creative Teaching Associates

COMPLETE THESE DOMINOES AND FIND THE SUM.

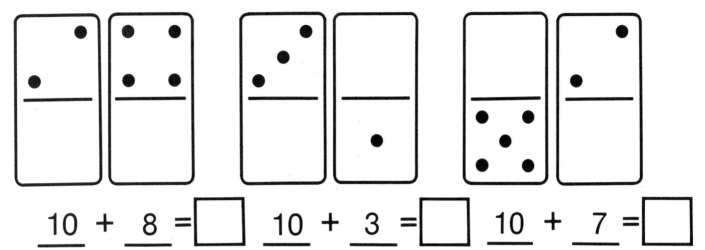

10 + 8 = ☐   10 + 3 = ☐   10 + 7 = ☐

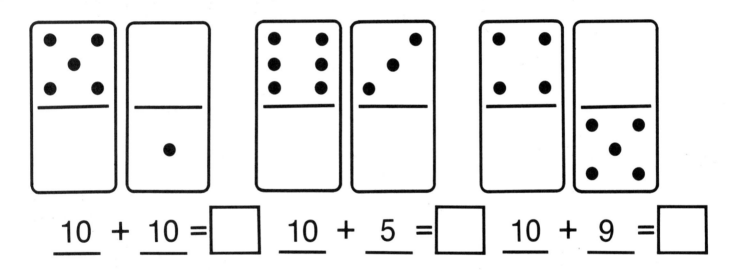

10 + 10 = ☐   10 + 5 = ☐   10 + 9 = ☐

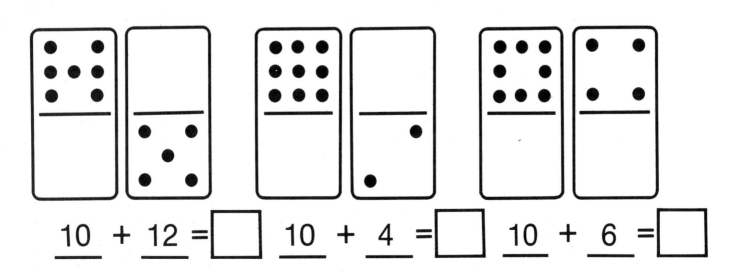

10 + 12 = ☐   10 + 4 = ☐   10 + 6 = ☐

© 1973 Creative Teaching Associates

COMPLETE THESE DOMINOES AND FIND THE SUM.

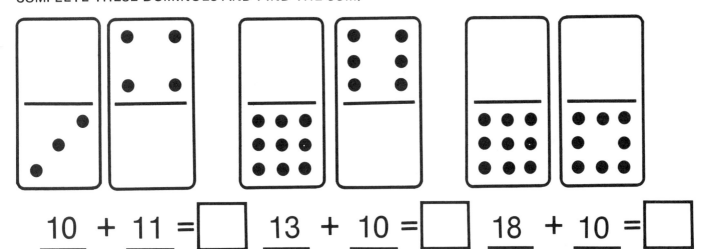

$\underline{10}$ + $\underline{11}$ = ☐ $\underline{13}$ + $\underline{10}$ = ☐ $\underline{18}$ + $\underline{10}$ = ☐

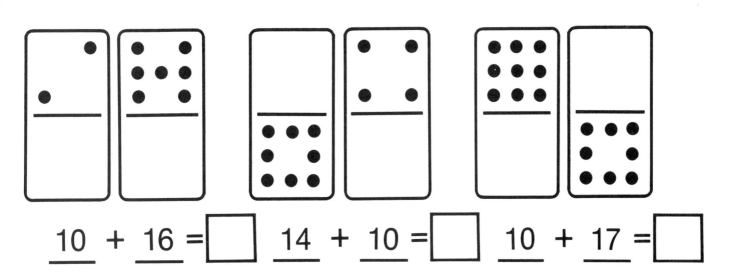

$\underline{10}$ + $\underline{16}$ = ☐ $\underline{14}$ + $\underline{10}$ = ☐ $\underline{10}$ + $\underline{17}$ = ☐

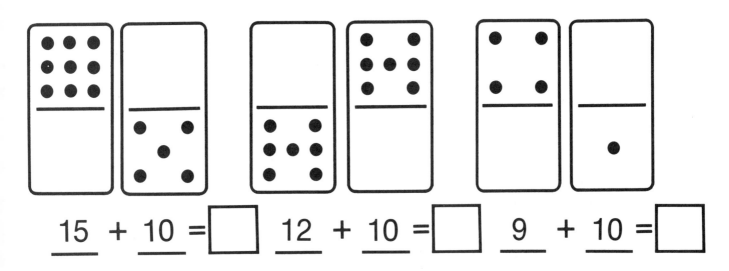

$\underline{15}$ + $\underline{10}$ = ☐ $\underline{12}$ + $\underline{10}$ = ☐ $\underline{9}$ + $\underline{10}$ = ☐

© 1973 Creative Teaching Associates

COMPLETE THESE DOMINOES.

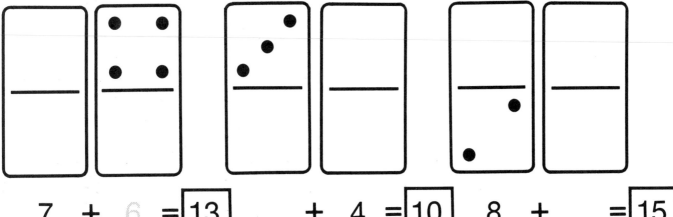

7 + 6 = 13   __ + 4 = 10   8 + __ = 15

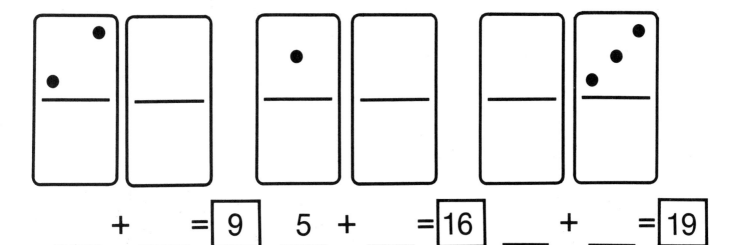

__ + __ = 9   5 + __ = 16   __ + __ = 19

6 + __ = 14   __ + __ = 17   __ + 11 = 20

© 1973 Creative Teaching Associates

FIND THE RIGHT DOMINOES.

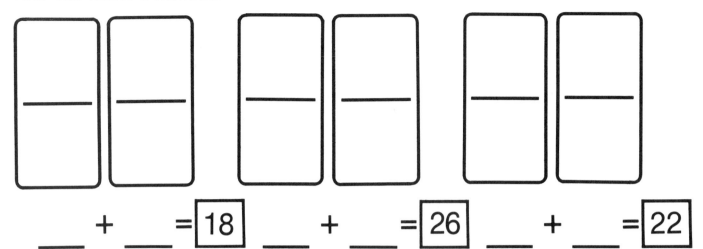

__ + __ = |18| __ + __ = |26| __ + __ = |22|

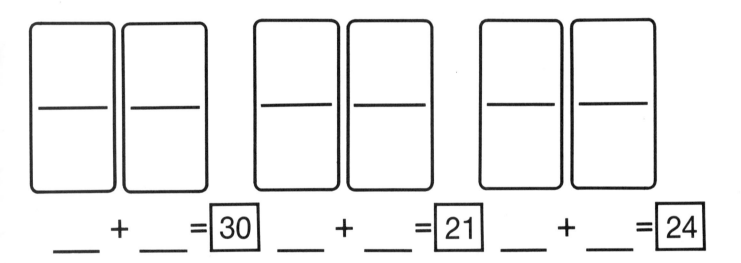

__ + __ = |30| __ + __ = |21| __ + __ = |24|

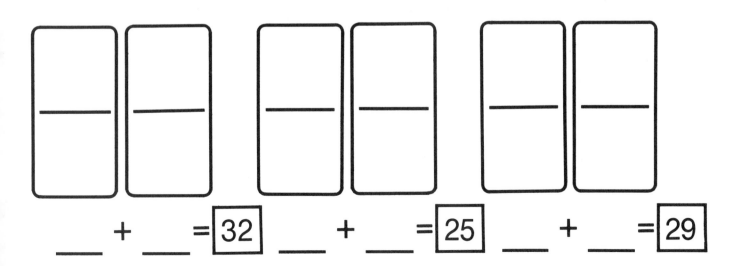

__ + __ = |32| __ + __ = |25| __ + __ = |29|

© 1973 Creative Teaching Associates

FIND THE RIGHT DOMINOES.

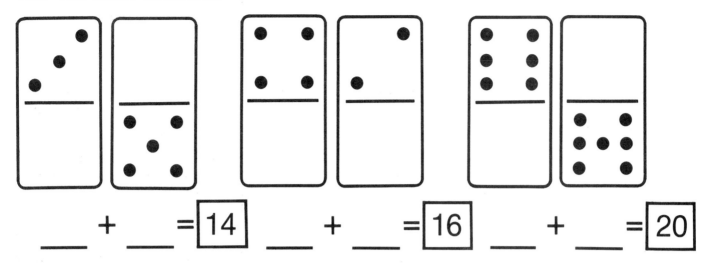

___ + ___ = 14   ___ + ___ = 16   ___ + ___ = 20

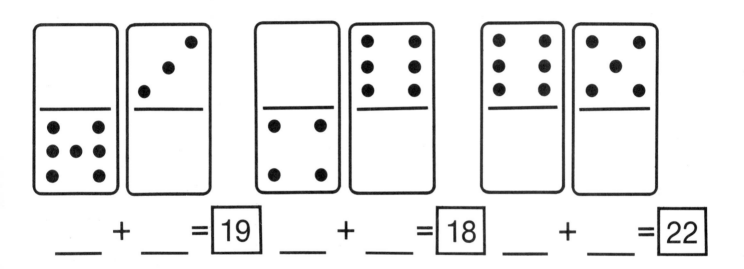

___ + ___ = 19   ___ + ___ = 18   ___ + ___ = 22

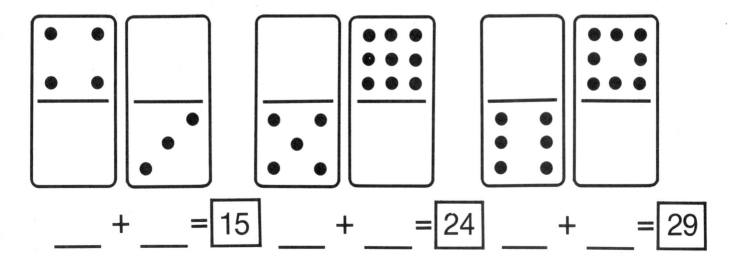

___ + ___ = 15   ___ + ___ = 24   ___ + ___ = 29

© 1973 Creative Teaching Associates

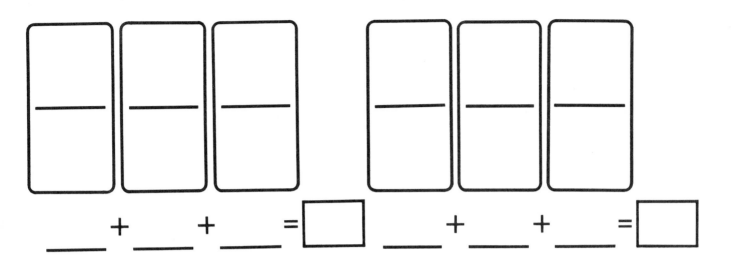

___ + ___ + ___ = ☐   ___ + ___ + ___ = ☐

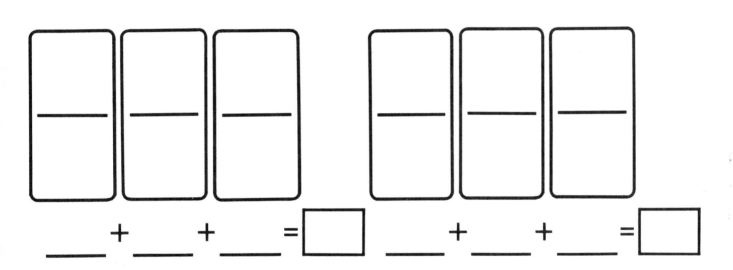

___ + ___ + ___ = ☐   ___ + ___ + ___ = ☐

___ + ___ + ___ = ☐   ___ + ___ + ___ = ☐

© 1973 Creative Teaching Associates

FIND ALL SUMS.

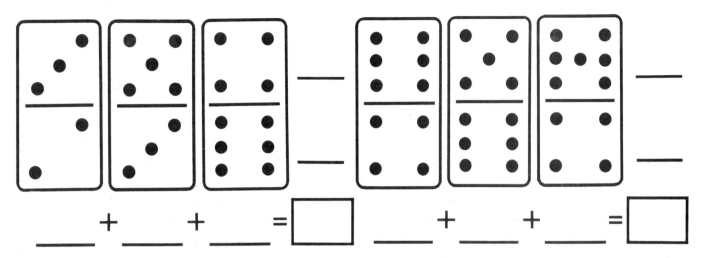

___ + ___ + ___ = ☐   ___ + ___ + ___ = ☐

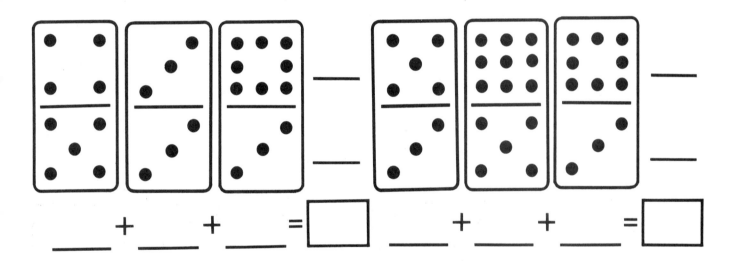

___ + ___ + ___ = ☐   ___ + ___ + ___ = ☐

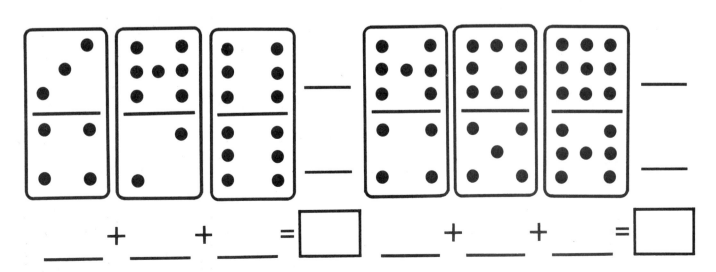

___ + ___ + ___ = ☐   ___ + ___ + ___ = ☐

© 1973 Creative Teaching Associates

FIND THE RIGHT DOMINOES AND ALL SUMS.

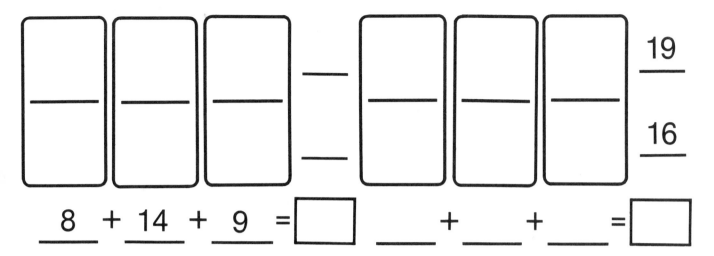

$\underline{\quad 8 \quad}$ + $\underline{\quad 14 \quad}$ + $\underline{\quad 9 \quad}$ = ☐ $\qquad$ ___ + ___ + ___ = ☐

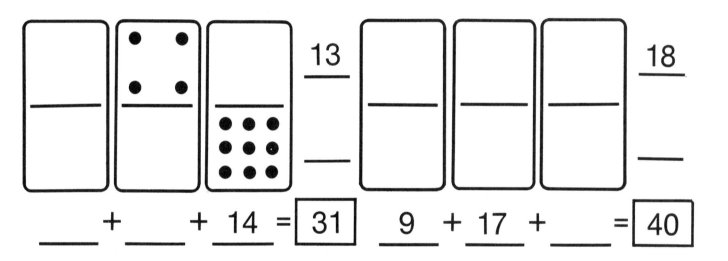

___ + ___ + 14 = $\boxed{31}$ $\qquad$ 9 + 17 + ___ = $\boxed{40}$

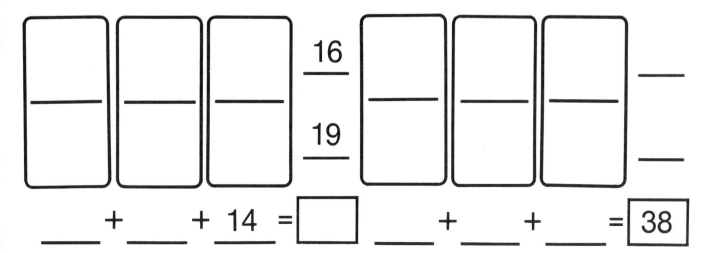

___ + ___ + 14 = ☐ $\qquad$ ___ + ___ + ___ = $\boxed{38}$

© 1973 Creative Teaching Associates

FIND DOMINOES THAT FIT THIS SUM. FIND ALL SUMS.

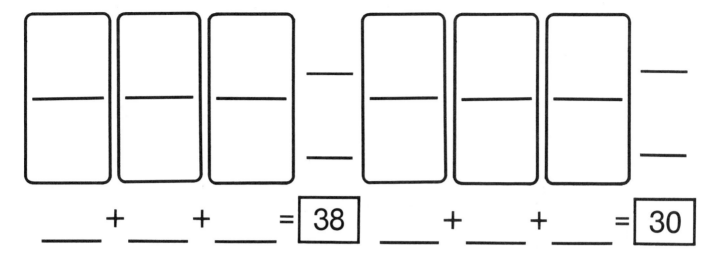

___ + ___ + ___ = 38   ___ + ___ + ___ = 30

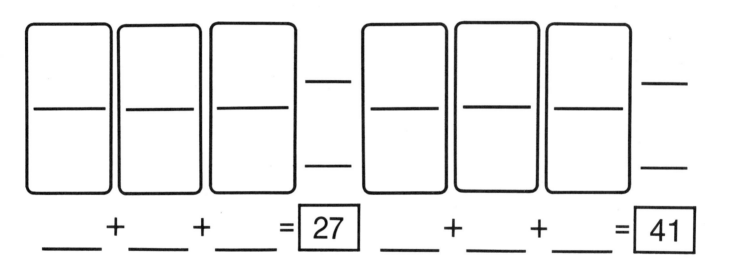

___ + ___ + ___ = 27   ___ + ___ + ___ = 41

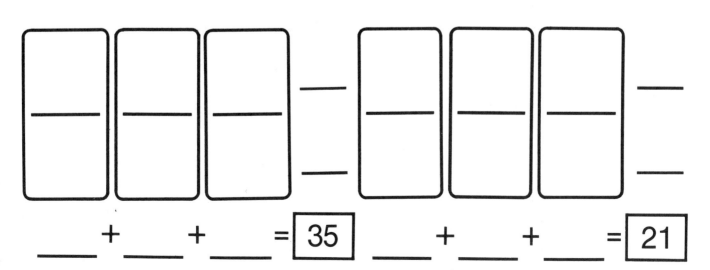

___ + ___ + ___ = 35   ___ + ___ + ___ = 21

© 1973 Creative Teaching Associates

NAME _____ ⟨38⟩

FIND ALL THE SUMS.

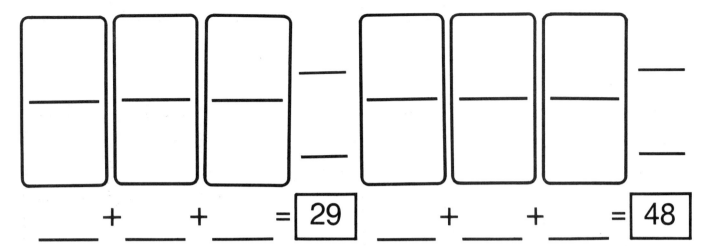

___ + ___ + ___ = 29   ___ + ___ + ___ = 48

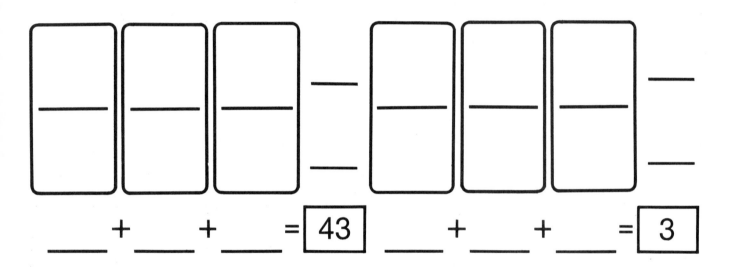

___ + ___ + ___ = 43   ___ + ___ + ___ = 3

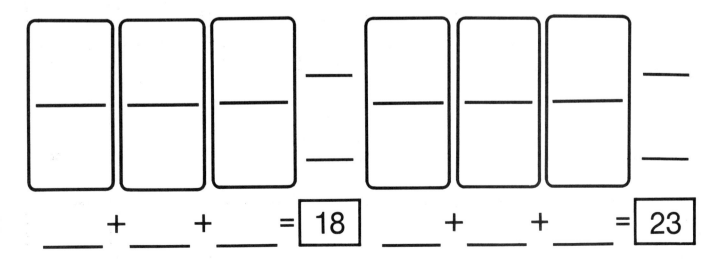

___ + ___ + ___ = 18   ___ + ___ + ___ = 23

© 1973 Creative Teaching Associates

FIND ALL SUMS.

___ + ___ + ___ = [ ]

___ + ___ + ___ = [ ]

___ + ___ + ___ = [ ]

___ + ___ + ___ = [ ]

© 1973 Creative Teaching Associates

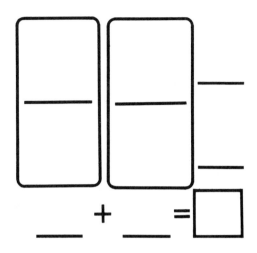

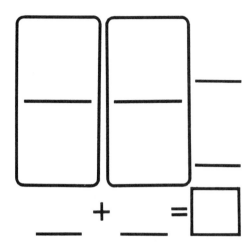

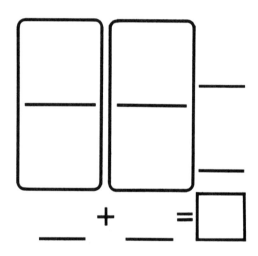

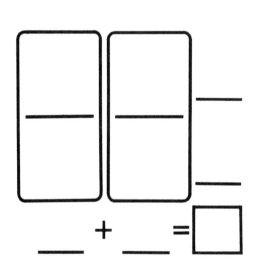

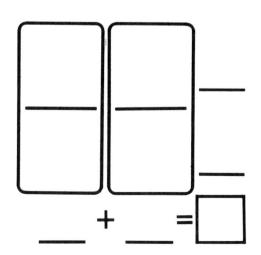

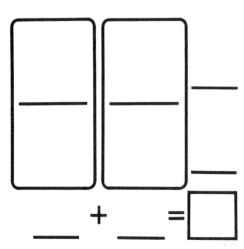

© 1973 Creative Teaching Associates

NAME _____

FIND ALL SUMS.

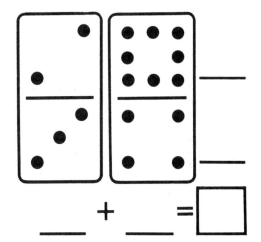

___ + ___ = [ ]

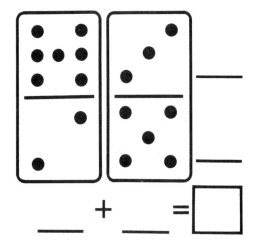

___ + ___ = [ ]

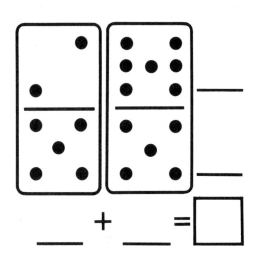

___ + ___ = [ ]

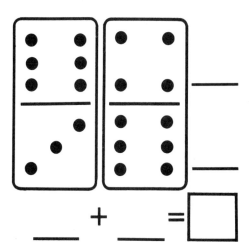

___ + ___ = [ ]

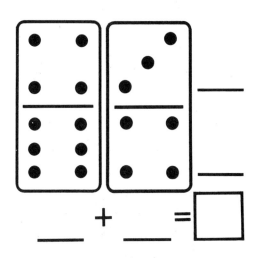

___ + ___ = [ ]

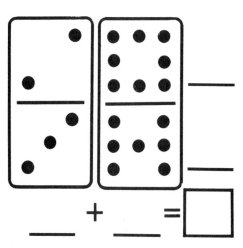

___ + ___ = [ ]

© 1973 Creative Teaching Associates

FIND ALL SUMS.

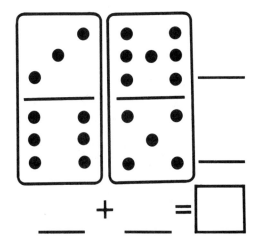

___ + ___ = ☐

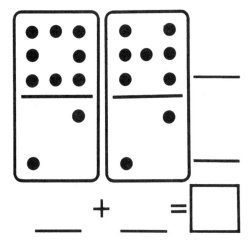

___ + ___ = ☐

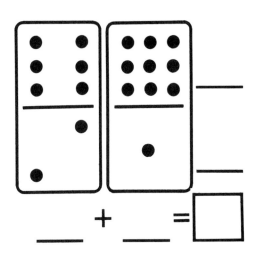

___ + ___ = ☐

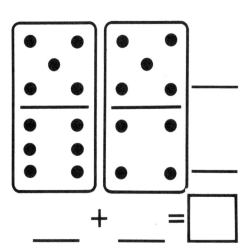

___ + ___ = ☐

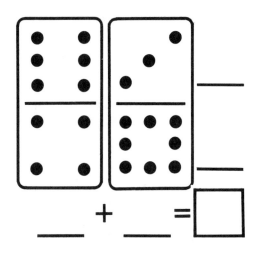

___ + ___ = ☐

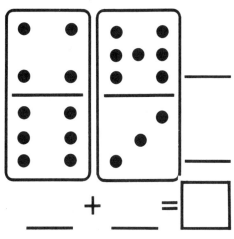

___ + ___ = ☐

© 1973 Creative Teaching Associates

FIND ALL SUMS.

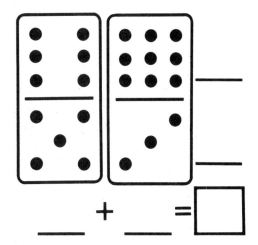

$$\underline{\quad} + \underline{\quad} = \boxed{\quad}$$

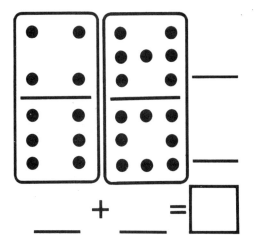

$$\underline{\quad} + \underline{\quad} = \boxed{\quad}$$

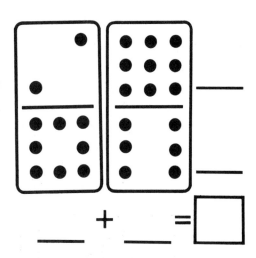

$$\underline{\quad} + \underline{\quad} = \boxed{\quad}$$

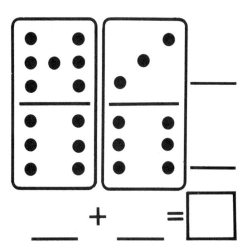

$$\underline{\quad} + \underline{\quad} = \boxed{\quad}$$

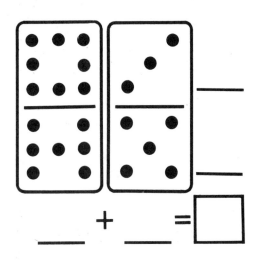

$$\underline{\quad} + \underline{\quad} = \boxed{\quad}$$

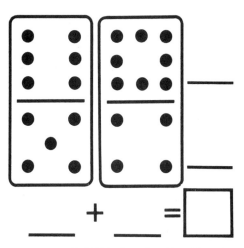

$$\underline{\quad} + \underline{\quad} = \boxed{\quad}$$

© 1973 Creative Teaching Associates

COMPLETE EACH DOMINO. FIND ALL SUMS.

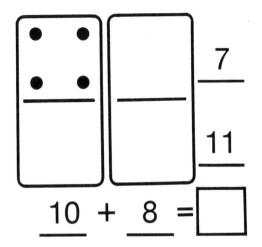

7
___

11
___

10 + 8 = ☐

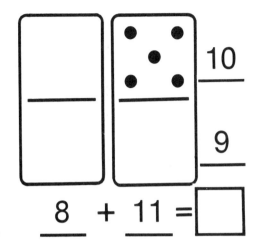

10
___

9
___

8 + 11 = ☐

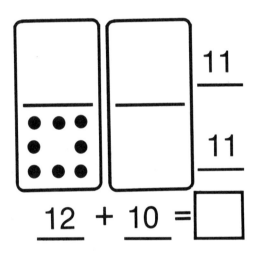

11
___

11
___

12 + 10 = ☐

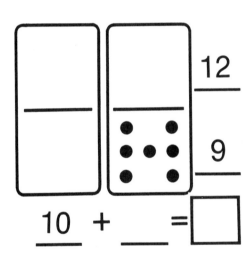

12
___

9
___

10 + ___ = ☐

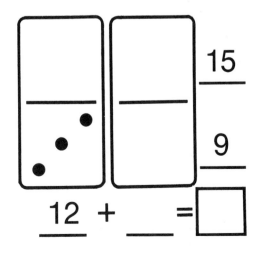

15
___

9
___

12 + ___ = ☐

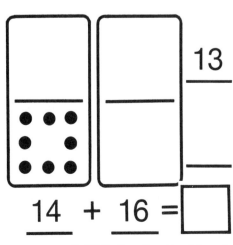

13
___

___

14 + 16 = ☐

© 1973 Creative Teaching Associates

COMPLETE EACH DOMINO. FIND ALL SUMS.

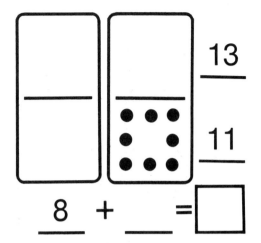

13
___

11
___

8 + ___ = ☐
___

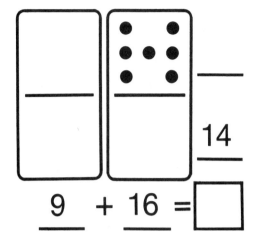

___

14
___

9 + 16 = ☐
___

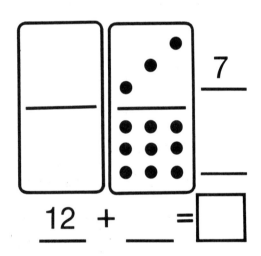

7
___

___

12 + ___ = ☐
___

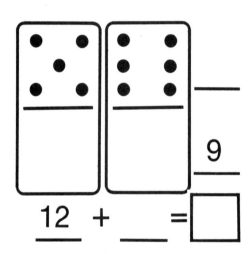

___

9
___

12 + ___ = ☐
___

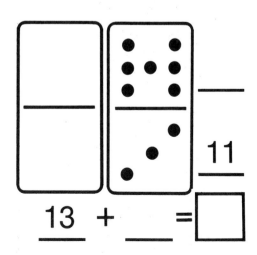

___

11
___

13 + ___ = ☐
___

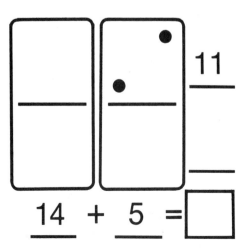

11
___

___

14 + 5 = ☐
___

© 1973 Creative Teaching Associates

NAME _____ (46)

COMPLETE EACH DOMINO. FIND ALL SUMS.

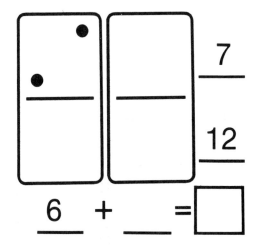

7
___

12
___

6 + ___ = ☐

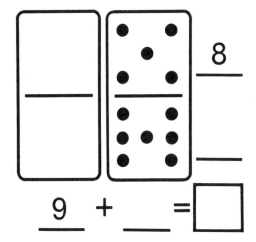

8
___

___

9 + ___ = ☐

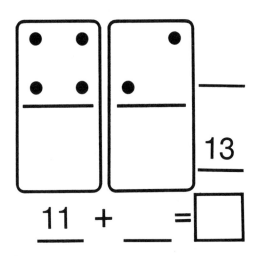

___

13
___

11 + ___ = ☐

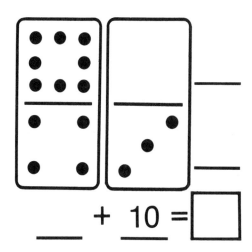

___

___

___ + 10 = ☐

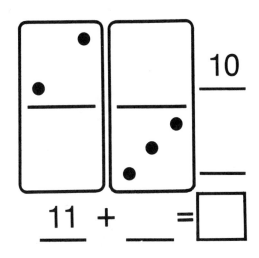

10
___

___

11 + ___ = ☐

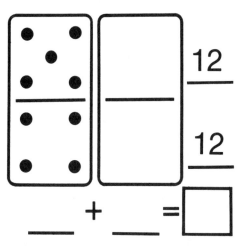

12
___

12
___

___ + ___ = ☐

© 1973 Creative Teaching Associates

FIND THE RIGHT DOMINOES. FIND ALL SUMS.

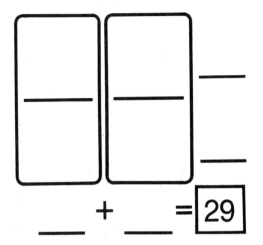

___ + ___ = $\boxed{29}$

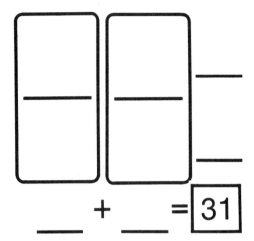

___ + ___ = $\boxed{31}$

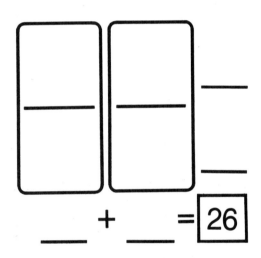

___ + ___ = $\boxed{26}$

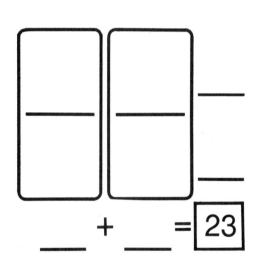

___ + ___ = $\boxed{23}$

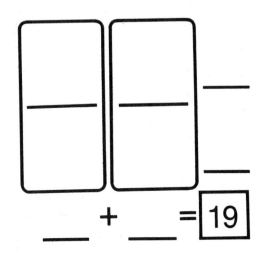

___ + ___ = $\boxed{19}$

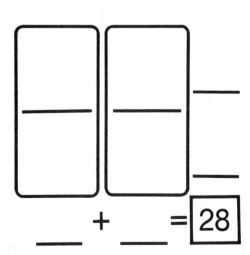

___ + ___ = $\boxed{28}$

© 1973 Creative Teaching Associates

FIND THE RIGHT DOMINOES. FIND ALL SUMS.

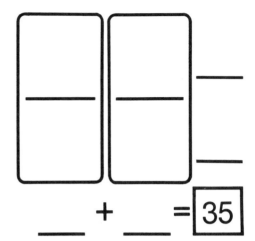

___ + ___ = 30

___ + ___ = 35

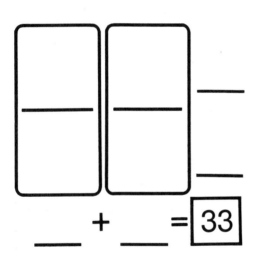

___ + ___ = 33

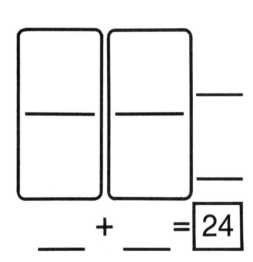

___ + ___ = 24

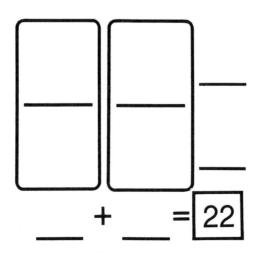

___ + ___ = 22

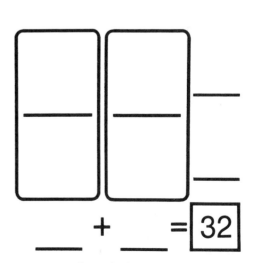

___ + ___ = 32

© 1973 Creative Teaching Associates

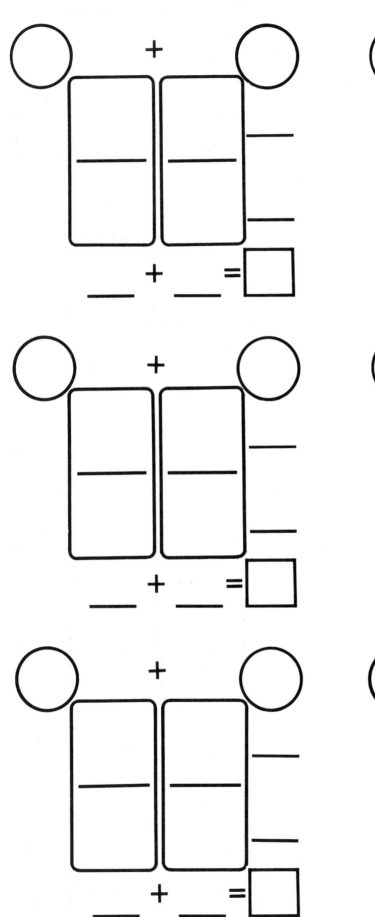
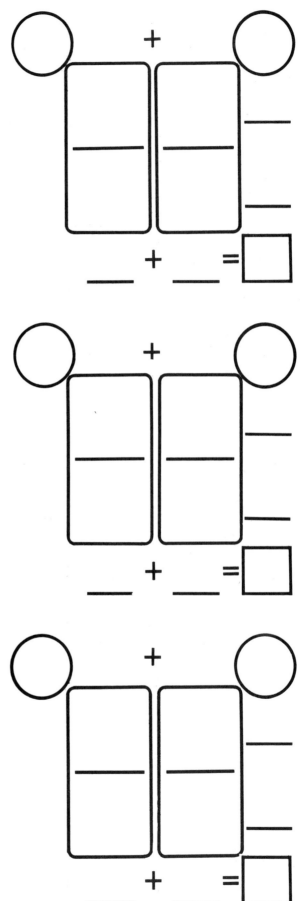

© 1973 Creative Teaching Associates

COMPLETE EACH DOMINO. FIND ALL SUMS.

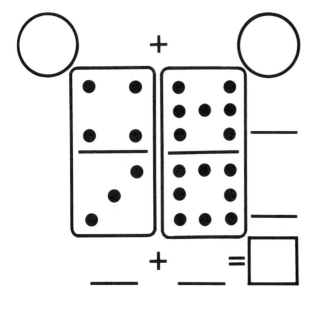

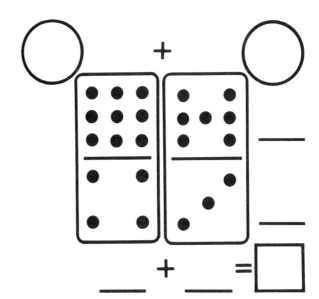

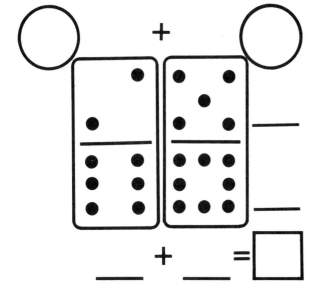

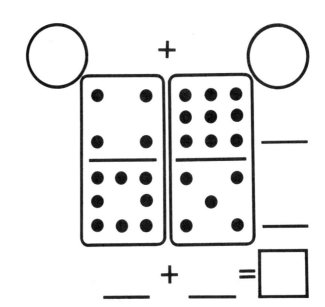

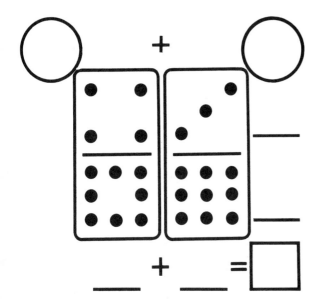

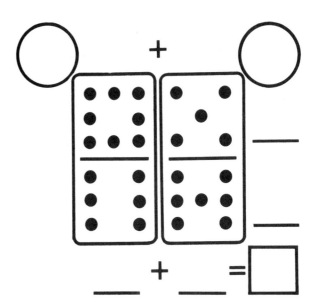

Creative Teaching Associates

FIND ALL SUMS.

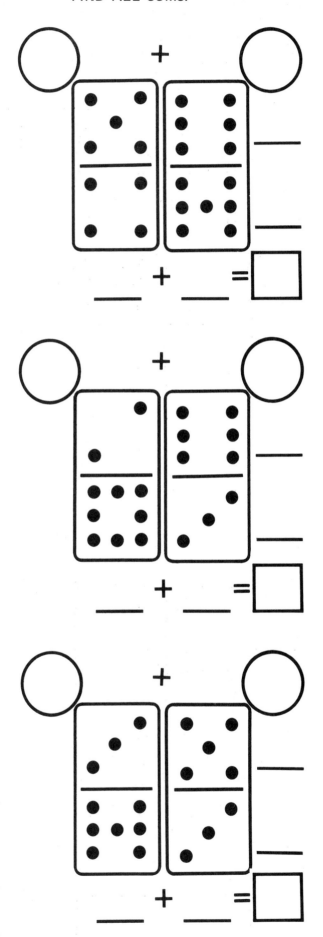
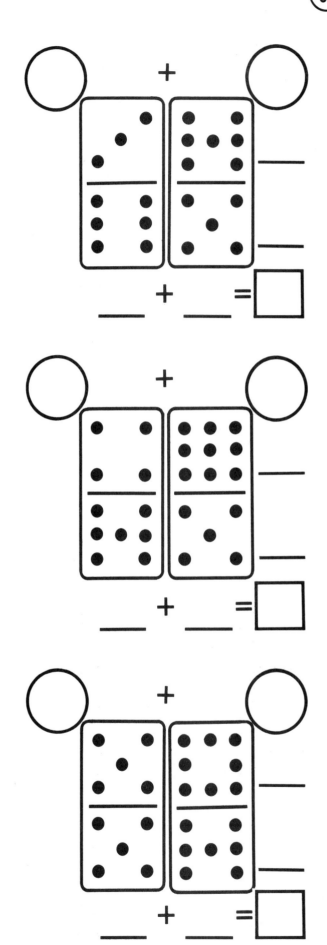

© 1973 Creative Teaching Associates

NAME _____

COMPLETE EACH DOMINO. FIND ALL SUMS.

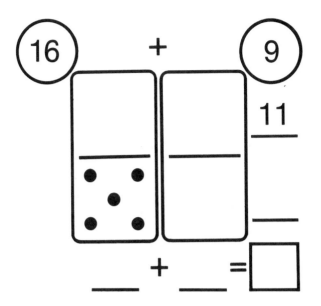

(16) + (9)

11
___

___
___ + ___ = ☐

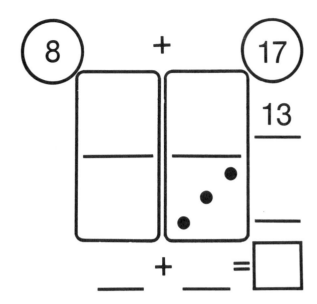

(8) + (17)

13
___

___
___ + ___ = ☐

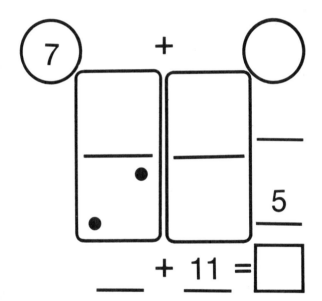

(7) +

___

5
___
___ + 11 = ☐

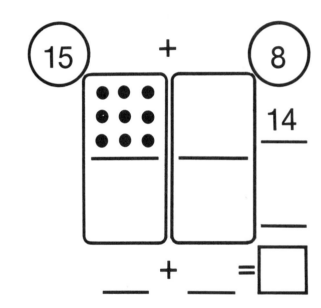

(15) + (8)

14
___

___
___ + ___ = ☐

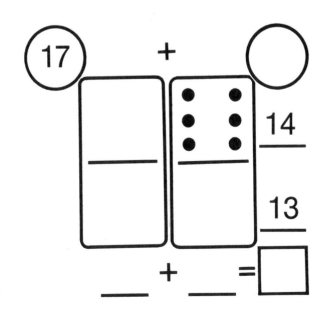

(17) +

14
___

13
___
___ + ___ = ☐

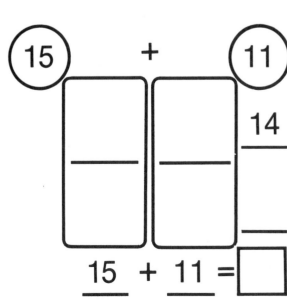

(15) + (11)

14
___

___
15 + 11 = ☐

© 1973 Creative Teaching Associates

COMPLETE EACH DOMINO. FIND ALL SUMS.

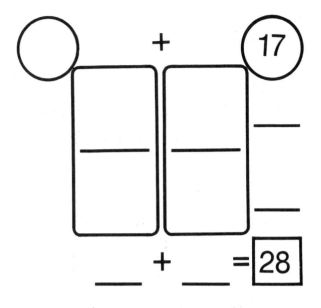

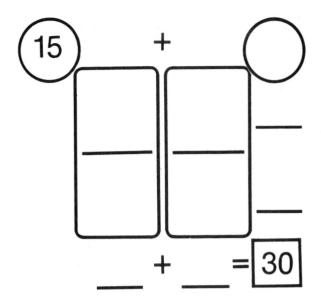

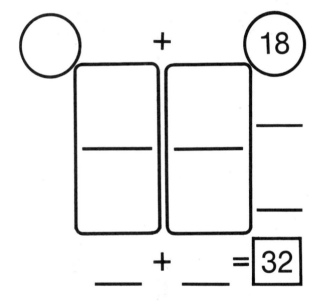

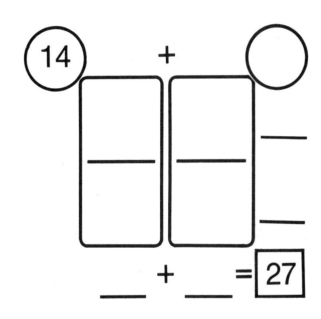

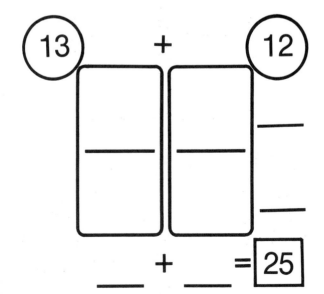

© 1973 Creative Teaching Associates

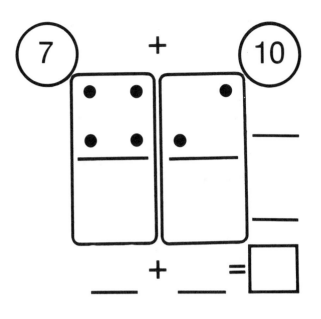

7 + 10

___ + ___ = ☐

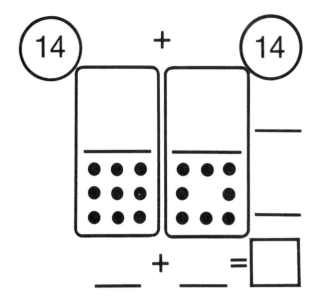

14 + 14

___ + ___ = ☐

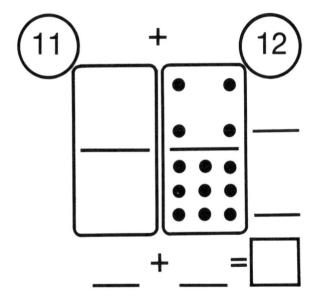

11 + 12

___ + ___ = ☐

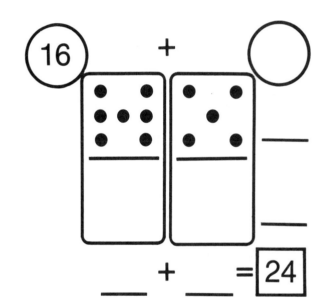

16 +

___ + ___ = 24

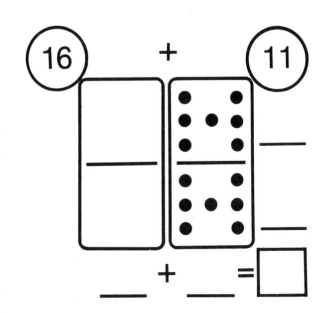

16 + 11

___ + ___ = ☐

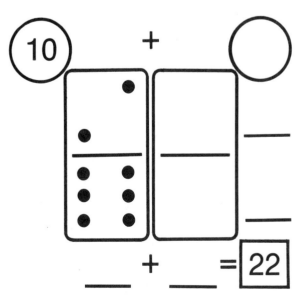

10 +

___ + ___ = 22

© 1973 Creative Teaching Associates

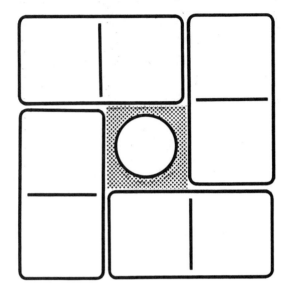

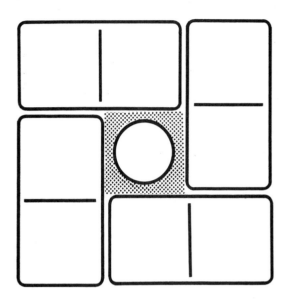

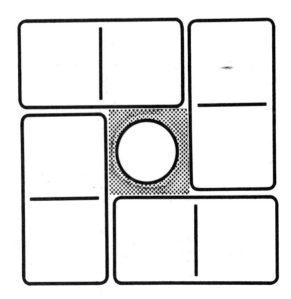

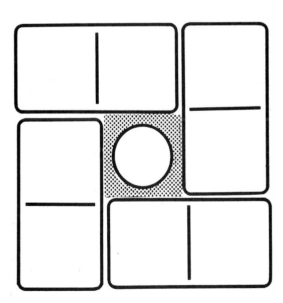

© 1973 Creative Teaching Associates

FIND DOMINOES IN EACH CASE SO THAT BOTH HORIZONTAL AND BOTH VERTICAL
ADDENDS HAVE THE SUM SHOWN IN THE CIRCLE.

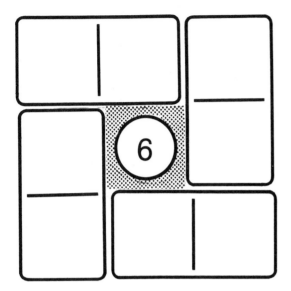

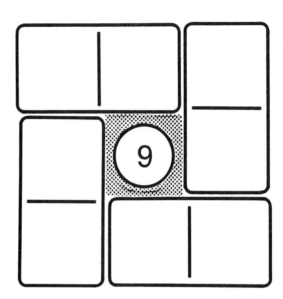

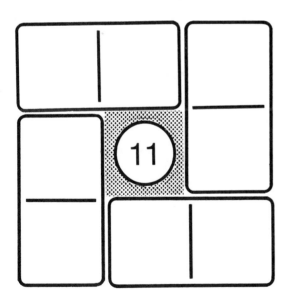

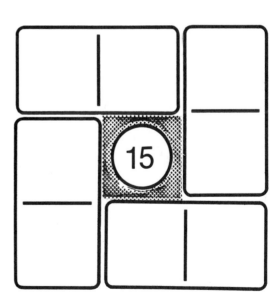

© 1973 Creative Teaching Associates

FIND THE RIGHT DOMINOES SO EACH SUM, VERTICAL AND HORIZONTAL, IS THE
SAME IN A PROBLEM.   WRITE THE SUM IN THE CIRCLE.

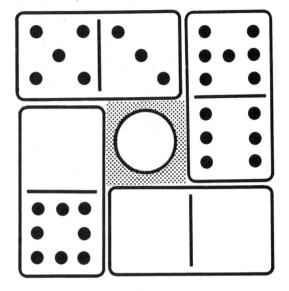

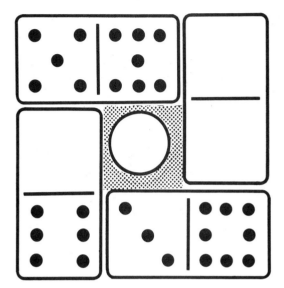

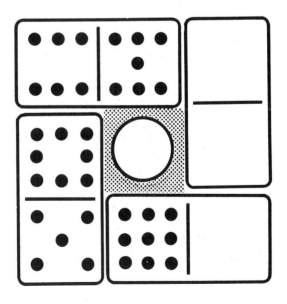

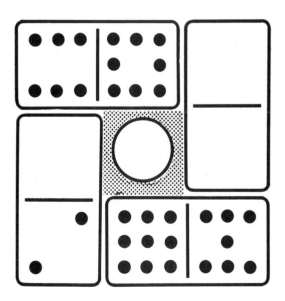

© 1973 Creative Teaching Associates

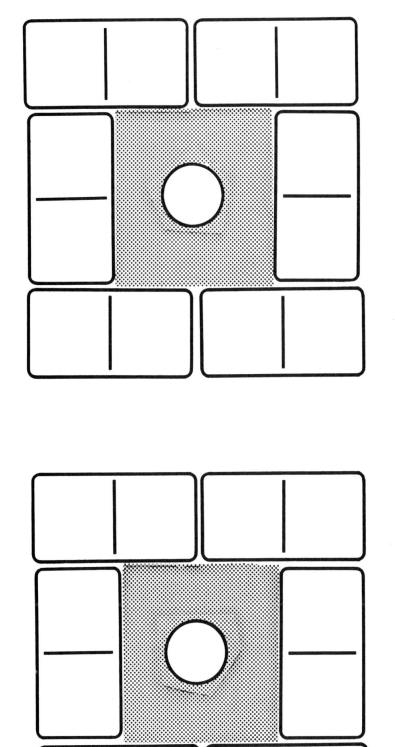

© 1973 Creative Teaching Associates

SELECT DOMINOES SO EACH COLUMN HAS THE SAME SUM.

© 1973 Creative Teaching Associates

SELECT DOMINOES SO EACH COLUMN HAS THE SAME SUM.

© 1973 Creative Teaching Associates